Many Thoughts On...

By Zeno Aurilius

Table of Contents

Intro

Dear Reader,

This novel covers many topics from science to politics to culture and art. It is written in short essays that are either argumentative or discursive. All sides are presented as a perspective on an issue and are not fact until proven beyond doubt.

Sincerely,
Zeno Aurelius

On Climate Change And "Green" Energy

From 1800 – 2000 the world's mean temperature has only increased by about 1.8 F. The media portrays a doomsday scenario for the world but really it is not very bad if at all. While some say that stricter environmental policies would reverse the damage we have caused, doing so would undermine freedom and possibly cause more harm not less. By intensifying the environmental policies of today we would be undermining what freedom stands for, the ability to do what is right. By doing this they would not allow us to choose what is right and act on it but, make us choose what they want, based on what they want. This would also possibly cause more harm than good because some renewable resources are harmful to us and the environment. One example would be the solar panels are radioactive and when they can no longer produce energy, they have to be buried 100 feet underground to slow the radiation and still some of it will reach the surface. While some say that climate change's damage is destroying our world, in reality it is not as harmful as they say.

Climate change is not a big of a problem as the media likes to portray. Patrick Moore the co-founder of Greenpeace explains this when he says that "No, significant warming in the 21st century." This quote shows that the climate change problem that the media says has been happening in recent years has not been happening. This shows that climate change is not destroying earth as much as most people would expect.

Next, the damage caused by climate change is not very harmful. The oxford database for death rates from floods, extreme temperatures, droughts, and storms shows that in the early 1900s there were 130 dead per million every year in the 2010s there was a 97% drop to less than 4 dead per million. This statistic shows that climate change related deaths are not increasing but decreasing due to the increase in better infrastructure. This shows that climate change is not very deadly.

Some say that solar and wind power are our future, but they actually do more harm than good. Alex Epstein says, "There is not one, real or proposed, independent free-standing, solar or wind plant." The reason behind this is simple, they both need back up of other energy sources which a main one would be fossil fuel. Wind plants need many materials including concrete and rare earth metals which include neodymium which would mean there would need to be an increase in mining which would then need fossil fuels to operate and create deforestation. Also, solar and wind plants have both been shown to kill many types of birds which include the threatened golden eagle.

In conclusion, climate change is not going to cause doomsday and is not very deadly. This is important because if continued, the climate alarmism will continue to cause an increase in debt and lead down a road of misinformation. If understood properly the future generations will not have to live in a world of fear.

References:

P. (2015, July 26). What They Haven't Told You about Climate Change. Retrieved October 08, 2020, from

https://www.prageru.com/video/what-they-havent-t
old-you-about-climate-change/

P. (2015, November 30). Is Climate Change Our
Biggest Problem? Retrieved October 08, 2020, from
https://www.prageru.com/video/is-climate-change-o
ur-biggest-problem/

P. (2015, October 19). Can We Rely on Wind and
Solar Energy? Retrieved October 08, 2020, from
https://www.prageru.com/video/can-we-rely-on-wind-
and-solar-energy/

P. (2016, April 18). Climate Change: What Do
Scientists Say? Retrieved October 08, 2020, from
https://www.prageru.com/video/climate-change-wha
t-do-scientists-say/

P. (2016, July 28). Climate Change: What's So
Alarming? Retrieved October 08, 2020, from
https://www.prageru.com/video/climate-change-w
hats-so-alarming/

On The Influence Of Technology On The Wealth Gap

Around the world all of us are influenced by technology. In some places around the world the advances in technology have increased the wealth gap between the wealthy and the poor. In other places of the world, like the United States of America and Japan, the wealth gap has decreased as the poor get richer and rise from poverty. While some argue that the wealth gap is decreasing, others argue that it is increasing.

The wealth gap is decreasing in some places due to innovations in science and technology which helps increase the amount of jobs. This is shown during an industrial revolution which has a significant increase in inventions and businesses. Also, if the amount of new innovations and businesses increases it in turn increases the amount of jobs there is, which allows more people access to high paying jobs. This allows the wealth gap to decrease due to an increase in more opportunities for everyone.

Furthermore, having more businesses increases the competition for other businesses to bring their prices on goods lower, which will allow more money to be in citizens pockets. Also, having more businesses may make other businesses increase their wages to make their job more appealing to citizens. On the other hand, some people think that it may help increase the wealth gap because more automated technology will make more businesses fire workers because automated technology will cost less and will not require wages. Also, an increase in technology possibly will make better goods which could put other businesses out of business which make them have to lay off their workers.

Furthermore, if technology were to increase to a place where workers were not required anymore in any job, it could make everyone lose their job. This is shown when computers got better, and NASA did not need humans to calculate trajectory or anything else. This shows that technology has already started to replace people, which has made them get fired. This makes the wealth gap between people increase.

In conclusion, some people in society think that technology has helped to increase the Wealth gap between people while some believe that it has decreased it. In some instances, technology has increased it, when people are replaced by it is one example. In other instances, it has decreased by increasing the amount of jobs. Finally, technology has decreased the wealth gap because there is more instances and evidence that shows that technology increases wealth.

On Immigration's Impact On The Economy

There are some that say that immigration negatively affects a country's economy, while others say that it positively affects it. Immigration can negatively affect a country's economy because the immigrants could come into a country and take jobs from the people that already live there. Others say that it positively helps it because they can help us make more products to sell for a profit.

Immigration negatively hurts a certain country's economy. This is shown when people from Mexico come to the United States, they could be willing to be hired at a lower price than someone who already lives here. This can take jobs from the people who have lived here for a longer period of time, they then might not be able to get another job and may cause them to go into poverty. This is just one way that immigration can negatively affect a country's economy.

Another way is that when those people come and take the jobs they may not have as much experience and may make lower quality products. This hurts the economy because, then the owners of the store may sell the items at a lower price than their previous one and then the business may go bankrupt because they are not able to keep up with their bills and will have to then fire their workers causing unemployment. Others say though that immigration may positively help the economy. This is shown when an immigrant from another country comes to the USA and fills a job spot that has not been filled yet. This helps the economy because then the business may be able to fill more orders and therefore make more money.

Also, another way immigration can positively help a country is that an immigrant may come here as a refugee and then may be able to work for less wages. This helps the economy because it allows the business to make more products to make more profit. Also, this can help the country because then they could get experience at one job and then get a better one and make more money and then some of that money goes back into the economy through taxes. This is another way in which a country may be positively affected by immigration.

There are many positives and negatives to immigration. Immigration can help a country by allowing more products to be made. Immigration can also hurt a country because it can take jobs away from people. All in all, immigration has many negatives that outweigh the benefits and therefore harm the economy more than it helps.

On The Divisive Effect Of Holidays And Days Of Celebrations

National celebrations and holidays have the power to divide and unite. There are some people who believe that celebrations and holidays divide people more than they unite them. Also, in other instances there are some people who believe that celebrations and holidays unite people more than they divide them. In my country celebrations unite people more than they divide them and in some cases celebrations and holidays divide people more than they unite them.

In America the celebrations and holidays bring Americans together on a common event. An example is on July 4th, most Americans come together to remember their country's birth and struggle for independence and freedom. This shows that in my country holidays and celebrations bring us together and unite us from a common history.

Another way holidays and celebrations unite people more than they divide them is in many places around America many people celebrate Christmas. Christmas has the power to bring people who have little to nothing in common together with others who celebrate and can make them happy. Some people say that holidays and celebrations can separate people more than they unite them. One example is how some people will not celebrate Columbus Day because of the possibility of violent history in Columbus's life and it can separate the people who do separate it from the people who don't.

One way holidays and celebrations divide people more than they unite them is on Thanksgiving. Thanksgiving was made by President Lincoln to help unite the country and was based on the meal the American Colonists and Natives ate with each other. This separates people because some people think that the colonization of America is based on slavery and harming of native people and think by not celebrating they are helping to not further the remembrance of violence. This is one way holidays and celebrations can separate people more than they unite them.

In conclusion, some people believe that holidays divide people more than they unite them like in the case of Thanksgiving and Columbus Day. Others believe that holidays unite people more than they separate them, like in the case of Christmas and the Fourth of July. Finally, given the evidence of many people coming together and uniting, holidays and celebrations unite people more than they unite them.

On If Shakespeare's Themes Are Still Relevant

William Shakespeare is one of the great writers of the past era and has taught many life lessons through his writing. There are many instances of this, for example, "Romeo and Juliet" shows rivalry in some parts of the play which are still shown today in many aspects of life. Some people today believe that these life lessons are still relevant today because the themes are timeless. Others believe that these themes are not relevant because the themes can no longer be applied to modern-day life.

Shakespeare's themes are still relevant and one example of this is the theme of Rivalry in "Romeo and Juliet". In "Romeo and Juliet" two families cannot stand one another for reasons they cannot remember and they will not let Romeo and Juliet marry which leads to their deaths. This theme is still shown in many aspects like how certain countries don't like each other or different ethnicities fighting for reasons far in the past.

This is one-way Shakesphere's themes are still relevant in today's society.

Another example of how Shakespeare's themes in his writing are still relevant is in his play "Hamlet". In "Hamlet" one of the main themes is Jealousy. You find Jealousy because the former King, Hamlet's father, was killed by his brother to gain the throne and his wife. You see Jealousy in the modern age in families and between friends when one person wants what someone else wants and they take it. This is one-way Shakesphere's themes are still relevant in today's day and age.

On the other hand, some of the themes are no longer relevant in today's day and age and you see one example of this in "Romeo and Juliet". In "Romeo and Juliet " one of the reasons Juliet's parents do not want Juliet to marry Romeo is because they have already picked out a husband for Juliet because that was the custom that some of the families followed in Shakespeare's time. In the modern age, many people do not follow these customs anymore and instead, people chose their husband or wife for themself and usually, the family has little to no influence on them. This is one-way

Shakesphere's themes are no longer relevant in today's day and age.

Another example of how Skaksphere's themes are no longer relevant is in the play "Romeo and Juliet" In the play "Romeo and Juliet" the two rival families are high up in social status and they have servants that tend to their needs and Shakesphere included this because in his time servitude was widely prevalent. In modern times servitude is not very prevalent because humanity has become more independent and are not needing servants as much especially for the general population which is usually middle or lower-class families. This is another theme that is no longer found in wide prevalence in today's society.

In conclusion, many themes found in Shakespeare's works can be relevant today like how the theme of Rivalry found in "Romeo and Juliet" can today be found between warring countries. Another theme that is found in Shakespeare's works that is still relevant is the theme of Jealousy found in the play "Hamlet" and you can still see this in families and friends if one person wants what another person owns. Some themes though you cannot find anymore in today's society is the theme of parents picking the wife or husband of the child and you no longer see this anymore because people have become more independent and are choosing who to marry for themselves. Another theme you cannot find in the world anymore is the theme of servitude and this can be found in the play "Hamlet", you no longer find this theme anymore because many people have become more independent and do not rely on others as much. All in all, most of the themes found in Shakesphes's writing are still relevant in today's society because his themes are still timeless.

On Quotes By Marcus Aurelius

"You have power over your mind – not outside events. Realize this, and you will find strength."

—— Marcus Aurelius, Meditations

"Dwell on the beauty of life. Watch the stars, and see yourself running with them."

—— Marcus Aurelius, Meditations

"The happiness of your life depends upon the quality of your thoughts."

—— Marcus Aurelius, Meditations

"Everything we hear is an opinion, not a fact. Everything we see is a perspective, not the truth."

—— Marcus Aurelius , Meditations

"Waste no more time arguing about what a good man should be. Be one."

—— Marcus Aurelius, <u>Meditations</u>

"If you are distressed by anything external, the pain is not due to the thing itself, but to your estimate of it; and this you have the power to revoke at any moment."

—— Marcus Aurelius, <u>Meditations</u>

"When you arise in the morning think of what a privilege it is to be alive, to think, to enjoy, to love ..."

—— Marcus Aurelius, <u>Meditations</u>

"The best revenge is to be unlike him who performed the injury."

—— Marcus Aurelius, <u>Meditations</u>

"Accept the things to which fate binds you, and love the people with whom fate brings you together, but do so with all your heart."

—— Marcus Aurelius, <u>Meditations</u>

"The soul becomes dyed with the colour of its thoughts."

—— Marcus Aurelius, <u>Meditations</u>

"It is not death that a man should fear, but he should fear never beginning to live."

—— Marcus Aurelius, Meditations

"Our life is what our thoughts make it."

—— Marcus Aurelius, Meditations

"Never let the future disturb you. You will meet it, if you have to, with the same weapons of reason which today arm you against the present."

—— Marcus Aurelius, Meditations

"If someone is able to show me that what I think or do is not right, I will happily change, for I seek the truth, by which no one was ever truly harmed. It is the person who continues in his self-deception and ignorance who is harmed."

—— Marcus Aurelius, Meditations

"If it is not right do not do it; if it is not true do not say it."

—— Marcus Aurelius, Meditation

On If Space Exploration Is The Most Important Economic Frontier

Space exploration has strong economic and ecological impacts on long-term human survival. Space exploration can lead to new discoveries, like new earth, that can help further the life span of the human species. Space exploration can also pull funding away from more important scientific advancements, like solving world hunger, and may harm people by not tending to these needs in time. Some may say that space exploration is the most important scientific frontier for human long-term survival while others say that it is not as important as some perceive it to be.

Space exploration is the most important scientific frontier for human survival because it can make humanity a multi-planetary species. Space exploration can help humanity find an alternate planet to settle so if something were to happen to earth humanity would have a place to go to where it can continue to thrive. This shows that space exploration is the most important scientific frontier for human long-term survival because without a second earth if a cataclysmic event were to affect the earth and we did not have a second earth to go to all humanity would be lost to the stars. This evidence shows that space exploration is the most important scientific frontier for human long-term survival.

Space exploration is not the most important scientific frontier for human survival because there are other more important issues to deal with, for example, world hunger. World hunger is more important than space exploration because world hunger impacts a larger audience than space exploration and it has a large death rate from people starving. Also, world hunger is more important because it is more urgent, this is so because only a select few are impacted by space explorations benefits while if we were to solve world hunger we can impact the world including those living in poverty, for example, some African nations, because we can bring the world food prices down which will increase wealth because less money is going to buying groceries. This is one way space exploration is not the most important economic frontier for long-term human survival.

Space exploration is the most important scientific frontier for human survival because advancements in space exploration can help all people on earth. One way advancements in space exploration can help all people on earth is because astronauts need heating and cooling devices depending on where they are and this can be applied to here on earth because there are people who live in extreme cold and extreme heat, like in Africa where it is very hot and arid, and this technology can help them survive in these climates longer and it can also help scientists go there and study and understand more about the earth and its climate. Another technology that can be applied to people on earth is alternate power sources and this can be applied to us on earth because there are power countries that cannot afford the current power sources and an alternate one can help them save money and at the same time get the energy to their citizens. This is one way that space exploration is the most important scientific frontier for human long-term survival.

Space exploration is not the most important scientific frontier for human survival because it can lead to economic crises that can lead to more debt and poverty.

Space exploration is not very cheap and if it does not succeed in its mission it can lead to losses in money. For example, in the United States of America, the government already has a 23 trillion dollar debt and some rocket programs can cost over 100 billion, and with this debt comes an increase in poverty. This evidence shows that space exploration is not the most important scientific frontier for human long-term survival.

In conclusion, space exploration has many positives and negatives and many people disagree on if it is the most important scientific frontier. Some of the arguments that say that it is the most important frontier are that it can positively impact humanity and increase wealth. While the opposition says that it can increase debt and is not as urgent as other things. Finally, given all the evidence, space exploration is the most important scientific frontier.

On The Qualities Of A Successful Government

Governments and governmental systems come in all shapes and sizes with some good and bad qualities. Some of the good and successful qualities can be freedom of speech and individual liberty. Also, some other successful qualities can be privacy and the right to bear arms. There are many qualities that lead to the success of a government and the happiness of its citizens.

One good quality of a successful government is the freedom of speech. The freedom of speech consists of the ability to say what you want and protest but this can also guarantee the freedom to not say anything if one chooses to and what to believe whatever they want to. A successful government needs this quality because one needs to be able to say what they want without the possibility of being punished because they can point out a flaw in the government and they can help fix the problem so a better system can be made. Also, a successful government needs this right because if the people think that they cannot say what they want to say then they might create an uprising like in China where there are a large number of protests because the government censors a lot of things and restricts their religious freedoms. This is one quality that makes up a successful government.

Another good quality of a successful government is individual liberty. Individual liberty can be defined as the ability to pick whatever job you want to try and get and the ability to go where you want. This is important because a person should be able to decide what job they want to try and get, you see how this is a good thing in the United States of America where the people can choose what jobs they want and this leads to people being able to work up to higher paid jobs and become wealthy. Also, this is important because if people are not able to go where they want they might feel oppressed and revolt against the government. This is another quality that makes up a successful government.

Another good quality that makes up a successful government is the right to bear arms. This quality can be described as the ability to get weapons to protect oneself from harm or oppression. This is an important quality because if the government becomes too oppressive then the people have a way to protect themselves and fight back against the government. Also, this is an important quality because if a citizen tries to harm another citizen then they could protect themselves from harm.This is another quality that makes up a successful government.

Another good quality that makes up a successful government is privacy. This can be described as the right for the people to not be watched and for them to do what they want in private without judgment. This is an important quality because without the citizens would become like those in the book "1984" by George Orwell, in which the people are watched everywhere and can be punished for anything that the government does not agree with and this would also deny the right of the Freedom Of Speech. This is another quality that makes up a successful government.

In conclusion, there are many qualities that make up a successful government that include the Freedom of Speech which allows the citizens to say or not say anything. Also, another important quality is individual liberty and this can be told as their ability to go where you want and pick what job you want. Also, another important quality is the right to bear arms and this can be described as the ability to get weapons to defend yourself. Also, another important quality is privacy which can be described as not being watched everywhere and the ability to do what you want in private. Finally, there are many other important qualities to a successful government but these are some of the most important.

On Terraforming And Colonization Of Other Planetary Objects

It is possible to create a second earth and colonize it. Through colonization and terraforming, humanity can create another Earth-like planet. There are some who believe that the human race should have the power to terraform planets and colonize them so all of humanity will not be wiped out by a cataclysmic event. Also, some believe that by terraforming and colonizing these other worlds we will be able to get many benefits from economically to advances in technology and medicine that we can use down on Earth to help humanity. There are others who believe that the human race should not terraform and colonize other planets because those planets may be host to alien life and by terraforming and colonizing them it may destroy that life. Humanity should have the power to colonize and terraform other planets.

Humanity should have the power to colonize other planets so we can extend extinction for the human race. Colonization can extend humanity's life, like what the article Mars Colonization: Beyond Getting There says, "...with shrinking biodiversity and depleting resources threatening the very survival of humans on this planet. Colonization of other planets could potentially increase the probability of our survival." This explains that one of the best options for long term human survival and protection against an extinction-level event from wiping out humanity is the colonization of other planetary objects. Colonization can protect humanity because, if Earth were to get hit by a meteor or another catastrophic event were to occur Earth was rendered lifeless then if humanity were to have a colony on Mars the human race could still survive there and hopefully one day sometime after that humanity could return to Earth to restart it's biosphere. A colony on earth and another planet or many other planets can postpone the life span of the human race as a whole for longer than what it would be if we were just on earth because the Earth is overdue for a catastrophic event to occur which means that it could happen at any time and if we were not prepared humanity would become extinct. The quote also shows that our own plant is failing and we may not have much time left to make a plan to survive as a species. If we don't prepare options for human survival and our own environment becomes unlivable humanity will perish with our planet unless we have a plan of escape. This is one reason humanity should have the power to colonize other planets.

Humanity should have the power to colonize other planets so we can make new advances in technology to further our knowledge and help those who need it on Earth. By colonizing other planets, we may be able to find new technology to help people here on earth in many aspects of life. You can see how this works when Mars Colonization: Beyond Getting There says, "As an example, Orwig points to the image analysis algorithm originally developed for extracting information from blurry images received from Hubble Space Telescope. After the technology was shared with a medical practitioner and as a result applied to medical images, such as X-ray images, it enabled more accurate visualization...". This shows that advancements because of space exploration are not only possible but, it happens and when it does it can impact and even save many lives of people on Earth. Also, by colonizing other planets we may be able to find new ways of exploring the different biomes around us. One way this may happen is if we colonize Mars we will have to find new ways for keeping the astronauts at a stable Earth-like temperature, this then can be applied to keeping Arctic and Antarctic scientists warm for long periods of time so they will not perish as quickly. Also, if we were to colonize Mars we would need to design new methods of growing food more resistant to the harsh elements so, the colonists will have a sustainable food source, this then can be applied to earth to feed the Earth's growing population and can help feed people in developing countries or regions where it is hard to grow food, for example in Africa it can be hard to grow food because of it's intense heat and mostly infertile ground. This is one reason humanity should have the power to colonize other planets.

Humanity should have the ability to and should colonize other planets so we can help increase jobs and boost the world's economy and average wealth. Colonizing other planets has better benefits than just new technology, if you were to colonize other planets and further space travel the revenue that is produced will be a large monetary amount and will provide many jobs as said by Space, the Final Economic Frontier,

"Both Neil deGrasse Tyson and Peter Diamandis have been given credit for stating that Earth's first trillionaire will be an asteroid-miner." This shows that a large amount of money can be found by traveling and colonizing space which can be used for more space travel advancement, can be put back into the economy to help it grow, and be used to fund missions to help poverty down on earth to better their circumstances. Also, you see that colonization can create many jobs when Space, the Final Economic Frontier says, "Jeff Bezos, whose fortune from Amazon has funded the innovative space startup Blue Origin, has long stated that the mission of his firm is "millions of people living and working in space." and "Elon Musk (2017), who founded SpaceX, has laid out plans to build a city of a million people on Mars within the next century." This shows that by colonizing and exploring further into space humanity will create jobs, for example, engineers, scientists, astronauts, managers, and many others, and jobs will only increase the further we go into space. The increase in jobs will also possibly increase the number of young adults picking going to college because these jobs have increased and they will be able to get into them easier which will increase average wealth across the world.

Humanity should have the power to terraform other planets so humanity can have a survival and escape plan in case the Earth's ecosystems are destroyed. Terraforming can be summed up best by what Terraforming: Engineering Imaginary Worlds says about it, "Terraforming involves processes aimed at adapting the environmental parameters of alien planets for habitation by Earthbound life, and it includes methods for modifying a planet's climate, atmosphere, topology, and ecology. Combining the Latin terra for 'earth' or 'land' with the gerund 'forming,' the term refers to '[t]he process of transforming a planet into one sufficiently similar to the earth to support terrestrial life' and is chiefly associated with sf discourse ('Terraforming, n.,' 2015)." If we were to terraform other worlds we would be able to create planets that can support humans and the ecosystems of Earth. Terraforming can take colonization one step further and allow humans to not only live on the surface of the planets in habitats but to walk and breathe outside without equipment. Terraforming other planets can also help us learn how to terraform our own planet. The knowledge of terraforming our own planet can serve two purposes, one is in case the ecosystem starts collapsing we can use terraforming knowledge to fix our ecosystem before it collapses and two, in case the Earth's biosphere is destroyed by some cataclysmic event, for example, a meteor, we can use terraforming technology to revert the earth to it's previous bio-status. Humanity should have the power to colonize and terraform other planets.

On the other hand, some say that terraforming and colonization can destroy other planets and possibly any alien life that lives there because the Martian life would not be able to survive in the conditions that humans would need for habitation. In retrospect, if enough research is done terraforming can be done safely and effectively. Also as Mars Colonization: Beyond Getting There says, "...our chances of discovering intelligent life in space are quite low..." This shows that the actual chances of any colonists finding life are not likely so there should not be much concern but, if they were able to find some then we would have to keep the ecology as close to what life needs as possible but, this is not likely to happen. One reason why this is not likely is that the conditions of Mars are so harsh that most known life could not survive on it; only the toughest single-celled extremophiles could possibly survive there.

In conclusion, colonization of other worlds is needed if we are to postpone a tragic extinction for humanity. Also, we need to invest more into colonization because it can give us new advances in medicine and technology that can expand our understanding of protection and healing and help those who need our help. Furthermore, not only will colonization create new technology but it will help boost our economy and bring in a large sum of money that can be used to help those in need and it can increase the average wealth of the world's people and decrease unemployment for all the world's citizens. Terraforming of other planetary bodies by allowing us to recreate or heal our environment in case of tragic destruction. Even though there are some who say that colonization and terraforming could destroy any alien life, the chances of finding any are very small and if there are some they are most likely not intelligent. Terraforming and colonizing other planets can be explored further if more research was to be gathered into how terraforming might be possible and how it might be possible quickly it would become a more feasible idea. Another way terraforming and colonizing other planets can be explored further if more research was to be gathered into how colonists would get to their planet of destination and survive in the harsh conditions of space for long periods of time, this would make the idea of space travel and human settlement on other planetary bodies a more desirable choice and will help solidify the claim that we can settle other planets. Finally, terraforming and colonization of other worlds is the best case option in the event that the Earth's ecosystem is destroyed and terraforming can give us new advances in science, technology, and medicine and increase the number of jobs there are which increases the average wealth across the globe which will help give those who need help the help they need.

PAK, C. (2016). Introduction: Terraforming: Engineering Imaginary Environments. In Terraforming: Ecopolitical Transformations and Environmentalism in Science Fiction (pp. 1-17). Liverpool: Liverpool University Press. Retrieved February 5, 2021, from http://www.jstor.org/stable/j.ctt1gpcb56.4

Weinzierl, M. (2018). Space, the Final Economic Frontier. The Journal of Economic Perspectives, 32(2),

173-192. Retrieved February 5, 2021, from
http://www.jstor.org/stable/26409430

Levchenko, I., Xu, S., Mazouffre, S., Keidar, M.,
Bazaka, K., Global Challenges 2019, 3, 1800062.
https://doi.org/10.1002/gch2.201800062

On De-extinction

Have you ever wanted a pet Dodo? Some say that we should bring back animals that are extinct and some say we should not and there are a few that are in between. Bringing back animals from extinction has the power to shape ecosystems. Some of the main perspectives are that we should have the power to bring back animals from extinction, that we should not have the power to bring back animals from extinction, and that we should have the power to bring back animals from extinction but up to a point because we should bring back animals that have recently died out but not dinosaurs. A point you need to know is: De-Extinction: The process in which you take an animal out of extinction.

First, the first main perspective is that we should bring animals out of extinction because we could restore some of the harmed ecosystems and so we could study what the ecosystems looked like far back in time. There are many people already working on bringing back animals like, International Union for the Conservation of Nature, Revive and Restore, and numerous others who have a background in animal conservation, biology, and genetic engineering. This is significant because it could allow numerous species to be brought back, restoring ecosystems that have been destroyed. This is shown when CBC says ""The idea is that we could take some of their genes to create new animals that have the traits of extinct species and then re-introduce them into habitats where extinct species once were, we could actually restore those ecosystems in some way" explains Wray." A strength for this perspective is that it can be backed up by many sources to prove its point. A weakness is that it could go the opposite way and further damage the ecosystem.

Next, the second main perspective is that we should not have the power to bring back animals from extinction. This argument is a strong argument because it has a lot of validity, but it has a weaker side because it has not been tested. Most of the people that agree with this side of the argument will say that will say, like Yale Environment 360 says, "Spending millions of dollars trying to de-extinct a few species will not compensate for the thousands of populations and species that have been lost due to human activities, to say nothing of restoring the natural functions of their former habitats." This is a valid point because the de-extinction process has the possibility of completely running the ecosystem and the animals that have adapted to those animals no longer being present may, then themselves become extinct. Also, many species every year are going extinct and it would take a few years just to get one back.

Lastly, the third main argument is that we should have the power to bring back animals from extinction but up to a point because we should bring back animals that have recently died out but not dinosaurs. This argument is a very strong argument because it has the potential to please all sides but it is weak because it is uncommon. Some people say, like what the National Library of Medicine says, "Several authors argue that de-extinction should only be applied to recently extinct species. This is an assertion that can be undercut with one example: extinction due to habitat loss. For species that have gone extinct due to habitat destruction or alteration, there is no justifiable reason to pursue de-extinction until the habitat has been restored." This explains that recently extinct animals are best for de-extinction because, if you were to go farther in the extinction timeline, the habitat that some animals would need would not be available at this moment in time.

In conclusion, there are many perspectives on if we should or if we should not go through the process of de-extinction. Some of the perspectives we have shown have very credible and strong points and resources. The first main point of view, that humans should have the power to bring back animals from extinction, is the most valid and has the most benefits that could come from it. The first perspective also has the potential to fix the earth's ecosystems and return ecosystems that could be destroyed back into their natural state. Finally, the main point will do the most good in the long run and is the one the human race should be trying to achieve.

Carlin, N. F., Wurman, I., & Zakim, T. (n.d.). [PDF].
Stanford Law.

Hoath, L. (2017, October 11). How scientists are
bringing extinct animals back to life | CBC Radio.
Retrieved December 05, 2020, from
https://www.cbc.ca/radio/thecurrent/the-current-for-october-11-2017-1.4348178/how-scientists-are-bringing-extinct-animals-back-to-life-1.4348231

Medicine, C. (2020). Myths about Cloning.
Retrieved December 05, 2020, from
https://www.fda.gov/animal-veterinary/animal-cloning/myths-about-cloning

Novak, B. (2018, November 13).
De-Extinction. Retrieved December 05, 2020,
from
https://www.ncbi.nlm.nih.gov/pmc/articles/PMC6265789/

Redford, K. H. (2018). Synthetic biology offers extraordinary opportunities and challenges for conservation (U.S. National Park Service). Retrieved December 05, 2020, from https://www.nps.gov/articles/parkscience31-1_redford.htm

Ehrlich, P., & Ehrlich, A. H. (2014, January 14). The Case Against De-Extinction: It's a Fascinating but Dumb Idea. Retrieved December 11, 2020, from https://e360.yale.edu/features/the_case_against_de-extinction_its_a_fascinating_but_dumb_idea

On Hawking's And Darwin's Mistakes

Most people have heard of Hawking's Theory of The Big Bang and Darwin and his Theory of Evolution but, there are some problems with them. In this series we will be looking at all of the problems.

The First Mistake: The Scientific Mistake:

1. The Law of Conservation of Energy/The First Law of Thermodynamics

The law of conservation of energy states that energy cannot be created nor destroyed only transformed. If energy cannot be created then how did all the universe come into existence from nothing. Nothing cannot create something, if there was pre-existing energy then the universe already started and then it did not start from the big bang.

2. The Law of Conservation of Mass

The Law of Conservation of Mass is like the law above and states that matter cannot be created nor destroyed but only transformed, this means that matter cannot come from nothing because it cannot be created.

3. Cell Theory: The 3rd Part

The third part of cell theory states that "The third part, which asserts that cells come from preexisting cells that have multiplied, was described by Rudolf Virchow in 1858, when he stated omnis cellula e cellula (all cells come from cells)." This is also proven by Francesco Redi when he disproved that maggots cannot spontaneously generate from rotting meat. This leaves the question that if cells come from only other cells then how did the first cells come from not cells?

4. The Theory of Relativity

The Theory of Relativity is in equation form $E = mc^2$. Many people don't know what this means though, it states that "Energy equals mass times the speed of light squared", this means that matter and energy are interchangeable and are equal. This also means that matter could not have been created by energy because they are one and the same. Also, this theory gives the idea that we live in a fourth dimensional world, the fourth dimension is time. This means that time and matter are part of the universe and one is present when the other is, in order from the Big Bang to explode and create matter and energy there would have to be some pre-existing form of matter, but due to the Theory of Relativity this is impossible because, when matter is present there is time and if there was time before the Big Bang, that means that the Big Bang must have not have been the beginning.

On Quotes By Halyard Twist

"The rose is a beautiful flower put together with a billion roses, planets, stars, asteroids, comets, and black holes in a symphony of wonder. The universe becomes an infinitely beautiful picture to all who come looking, but many still choose to see the rose."
— Halyard Twist

"The one question no person can answer is: why?"
— Halyard Twist

"Every person is a genius, but when we start measuring that genius with letters, we let it wither. We say that those who have a certain letter will excel but those who have the letter

rarely make history while those who don't
make history."
— Halyard Twist

"The Yale graduate reads
history; the innovator makes it."
— Halyard Twist

"When we teach children that the letter A is
more valuable than the letter B, we have lost
our sense of worth, and when we say that + is
greater than -, we have lost our sense of
identity."
— Halyard Twist

"Some say fiction is an escape from reality,
when it is really a window to another one."
— Halyard Twist

"Non-fiction is a door, fiction is a window."
— Halyard Twist

"You can learn more from fiction than you could ever from non-fiction."
— Halyard Twist

"Paper is the gateway into the mind."
— Halyard Twist

"Fiction is the only place where everything is possible, and nothing is possible."
— Halyard Twist

"A good suspense story takes all the facts then dangles them just where the reader can't get them, then gives the reader one just every so often, like a carrot on a stick."
— Halyard Twist

"The test always comes before the grade."
— Halyard Twist

"Genius is not the ability to pass the
test, but for you to write the test."
— Halyard Twist

"The only thing that can transcend time and
space and can cross galaxies no matter how vast
the distance is friendship."
— Halyard Twist

"Every writer knows two languages: their
home language and imagination."
— Halyard Twist

On Social Media Censorship

Censorship on social media is like wearing a muzzle in public. If we don't stop censorship our country will fall as the first amendment is taken away. Freedom and freedom are the same if one goes, they all go. What you post on social media should not be censored.

Censorship is running repaint across our great country. If we don't stop censorship our country will fall as the first amendment is taken away. People want to censor speech on social media because they say it is "Offensive" If we fail to save free speech in America, we lose it globally.

Free speech in America creates hope for free speech globally. In other nations censorship is mandatory and if you don't follow their speech codes it can be punishable by extreme charges. One example is when The Fire says, "...in Pakistan, people are arrested and sentenced to death for "blasphemy" for insulting

Islam...". Another example is when The Fire says "...or France, where a man was fined for holding up a sign saying "Get lost, jerk" to French President Nicolas Sarkozy — words Sarkozy himself said to a critic who refused to shake his hand during a public event." These are just a few examples of why free speech is essential to all countries, not just the United States of America.

Free Speech in America is freedom for all. To create a law prohibiting the censorship of speech no matter if it is offensive or not across all platforms. This solution will allow people to say how they feel without fear of it being taken down. Some may say that speech can be violent but, as said in the examples above it is the limiting of free speech that is violent. American history is a great example of how free speech has freed us. Japan is an example of how free speech and freedom can save people and boost an economy.

Freedom is the one peg holding the entire idea of a free world together. Free speech is the hope for other nations to become free nations. If free speech goes, all freedom will go down with it too.
Censorship. (n.d.). Middle Tennessee State University | Middle Tennessee State University.

https://www.mtsu.edu/first-amendment/article/896/censorship

Free speech on campus is under attack. (2020, April 15). Speech First - Free speech on campus matters. https://speechfirst.org/?gclid=CjwKCAjwh7H7BRBBEiwAPXjadgm2E4xBEiyJNr2YLXZDrL0jSDHBy3X7vwyEKYCA4ZleVHX

Nunziato, D. C. (2011). How (Not) to Censor: Procedural First Amendment Values and Internet Censorship Worldwide. Scholarly Commons | George Washington University Law School Research. https://scholarship.law.gwu.edu/cgi/viewcontent.cgi?article=2540&context=faculty_publications

What to do about the emerging threat of censorship creep on the internet. (2017, November 20). Cato Institute. https://www.cato.org/publications/policy-analysis/what-do-about-emerging-threat-censorship-creep-internet?gclid=CjwKCAjwh7H7BRBBEiwAPXjadozXNkRflxDzTqPsso-HvD0ffzOHzs4MHu1TPiG6VPc3uP2xE4RfrRoCBpwQAvD_BwE

A world without hate speech. (2019, April 19). FIRE. https://www.thefire.org/a-world-without-hate-speech/?gclid=CjwKCAjwh7H7BRBBEiwAPXjadgi0qJHLtuOJ6X14fLYtlkl2QaR9LnO7HoxaKBLGOUrJN9yjB-_LERoCcvoQAvD_BwE

On Censorship

The freedom of speech is the base of all freedoms. There should not be consequences for what someone posts because the first amendment protects free speech. Others would say that what people write might be mean. People are protected from punishment of what they say about things by the first amendment, that includes social media. The first amendment holds all the amendments together if it is removed all the amendments will crumble. People should not be punished for what they say on social media even though it may be considered mean to some.

Some may say that social media is a privately-owned business the government can do anything with privately owned businesses. However, John Samples's article says, "...the court ruled that a company town, like a government, could not restrict First Amendment rights...". Others might also say that they should stop some social media users before something bad happens. On the contrary, "We just have felt like muzzling people's rights to communicate is not proper. What we can do instead is focus on sharing with them what the consequences are if they act inappropriately,"said the article by Luke Meredith and Michael Marot.

Next, there should not be consequences for what someone posts because the first amendment protects free speech. The freedom of speech protects those who say things. Furthermore, "The First Amendment offers strong protections against such restrictions." John Samples. Also, No one is allowed to restrict the first amendment even though it is a privately owned "place". Also, John Samples says, "...the court ruled that a company town, like a government, could not restrict First Amendment rights...".

Lastly, the first amendment protects those anywhere, even on social media. Putting prohibitions is destroying the first amendment. If people don't save the first amendment our country will fail. If they limit the first amendment, then all the amendments will crumble. In conclusion if people don't fight for what is right then we will not be able to fight anymore.

Samples, John. "Why the Government Should Not Regulate Content Moderation of Social Media." Cato Institute, 19 Apr. 2019, www.cato.org/publications/policy-analysis/why-gover nment-should-not-regulate-content-moderation-social -media

Meredith, Luke and Marot, Michael. "Software Helps School Monitor Athletes' Postings" eCampus News, 14 June, 2010 https://www.ecampusnews.com/2010/06/14/softwa re-helps-schools-monitor-athletes-postings/2/ https://flvsft912.flvs.net/webdav/educator_english1_ v18/module03/journey/pop/03_04_02a.htm

Brownstein, B. (2019, November 04). Without Free Speech, All Speech Becomes Government Speech: Barry Brownstein. Retrieved October 28, 2020, from https://fee.org/articles/without-free-speech-all-speech -becomes-government-speech/

On Columbus Day

I agree, the Natives do deserve a day (but for only the ones that did help). They did help, but the crew of his ships were on the edge of a mutiny before he was able to talk them down. If it were up to them, the New World may have never been found. Also, I am not saying slavery is good, but that is what they did back then. Also, many native tribes had slavery, and the earth has had slavery since the dawn of time; He did not invent it. Also, we should not have holidays for entire groups of people, instead we should have holidays for the people who did great things for humanity, and Columbus did a great thing for all of humanity. Also, Columbus once said, "Let those who are fond of blaming and finding fault, while they sit safely at home, ask, 'Why did you not do thus and so?'" Additionally, "To those who judge, put yourself in their place and do what you would do, then judge yourself by the standards you judge them both now and if you were them." - Zeno Aurelius and Matthew 7: 1-2 - "Do not judge, or you too will be judged. For in the same way you judge others, you will be judged, and with the measure you use, it will be measured to you.

The Egyptians had slaves, and we don't get mad at Egyptians. Some of the native tribes had slaves, and some did things that were worse than slavery, and we still have a day about them. Also, no person is perfect. Just because he did have some faults does not mean we should disrespect his name. Every person in history has done bad things, and we don't take their name out of the history books.

One example is that soldiers kill people to protect their country, and we don't disrespect their name. Also, some of the natives did horrible things which include slavery and even cannibalism. The Egyptians did horrible things but they gave incredible advances to the world and were the greatest architects that ever lived. When I say that the natives did bad things, I mean that they did bad things and still have a day that is honored (Columbus day is considered Indigenous Peoples day in some parts of America). In addition, he was peaceful with some of the tribes. In the beginning of when he got there, they both were interested in both of their cultures.

I also think that we should have Columbus day not to honor his faults, but to honor his accomplishments, like how he was able to connect the New World and the old one and sail across the

Atlantic when the boats of his day should not have. Also, if he were alive today, he probably would regret the things he did and ask for forgiveness so, yes, both the natives and Columbus had their faults but I think we should forgive them both and leave the past in the past. Also, we celebrate other people for discovering things when they were already there.

PragerU. (2018, October 7). *Goodbye, Columbus Day*. PragerU. Retrieved July 6, 2023, from https://www.prageru.com/video/goodbye-columbus-day/

PragerU. (2020, October 5). *Celebrating Columbus*. PragerU. Retrieved July 6, 2023, from https://www.prageru.com/video/celebrating-columbus/

Preview of "The Lost Forest"

This is a sneak peak for the book "The Lost Forest" by Halyard Twist coming Late 2024.

Synopsis:

Alexander and Ary find a hidden cave that leads them to a forest seemingly untouched by man. What will they find? The story is led by three perspectives, Alexander and Ary the Top Dwellers and Ji.

Entry #1: Alexander Kite

The date is Jan. 27, 2021 and I got this journal for Christmas to, as my parents said, "Write my adventures" I don't know what adventures they could be talking about since I have never had any kind of adventure that I ever wrote about until it was to late and I forgot to much. I never got to writing in it though because I did not know what to write but, I decided that I should probably start writing in it sometime. I should probably introduce myself though, my name is Alexander Kite and I am 15 years old and live in Cactus Lake, not an actual lake but a city name after the lake that used to be here. I have not seen a lake here though or in any of the surrounding area for that matter probably because, I live in the middle of a desert. I guess that's a given. There are not many people that would choose to live here, probably because the lack of cheap water and the sweltering heat. The school here is no better, it has no air conditioning, they say there is but I doubt it, though I don't go there much since I'm homeschooled and only go there for the horrible end of year tests. Not many people speak to me, I take that back not many people speak to anyone out here(especially this one kid, I think his name is James, he does not speak to any one and no one tries to speak to him, probably because, he keeps talking about some Medieval, so everyone thinks he is crazy). Well except my best friend, Ary Franklin, her real name is Aurora Franklin, but no one calls her that except her parents and even that is rare.

We met one day when we were both in sixth grade and we sat next to each other at lunch, she's homeschooled too. I had a tuna sandwich, but I don't like tuna very much, she had a ham sandwich hat she did not like but, she loved tuna. She asked if I wanted to switch and the rest is history. since then we always explored the mountains around Cactus Lake. We have mapped out some of it. The mountains are very interesting because our entire city is sounded by it in a 10 mile radius circle from the center. The mountains themselves are about a mile thick and 10,000 feet high. Tomorrow we are supposed to explore a section we have never explored yet, from up above it looks like it could be a clearing in a circle with about a 100 foot radius. I'll fill you in tomorrow on how it went. Wait, I just realized that I don't know who I am writing to, maybe the future me or maybe some person in the future interested in my life

I guess I see then.

Until later,

Alexander Kite

P.S. I go by Alex.

(Do you have to write a P.S. in a journal?)

Entry #2 Alex: The Cave

The date is Jan. 28, 20201

Ary and I are meeting up at our usual spot near town
hall where we met for the first time. We each have our
usual supplies which is, our backpack full of a
compass, this Journal, pens and pencils, flint, and
rope. We head out at about high noon and we walk
east towards the possible clearing in the mountains.
We check our map of the mountains we have made
so far and trek into journeyman's valley, this is the
only way in and out of our city by foot and we walk
for about an hour to the center then when we are
sure no one is watching, we do this so no one knows
our hideouts, we climb up the rock face towards the
clearing.
We eventually reached it after another hour. and we
climb down about 10 feet towards some trees. We
find out that it is definitely a clearing. We wander
around the beautiful landscape for a few minutes
when we hear it. It sounds like a waterfall, but that
can't be possible, we're in a desert. Confused, we
make our way towards it and by surprise find a huge
200 foot waterfall in front of us. It gleams with water
bouncing off the jet black rocks. We decide to walk

towards it and to investigate more when we notice a small crevice in the side of it. We look in the crevice and find that it is a 10 foot tall opening to what looks like a cave but we cannot see the end of it. We decided to walk into it just a little. I'll tell you more when I know more, but you gotta keep reading, deal?

Sincerely, Alex

P.S. I forgot to mention that we have flashlights and our map we have drawn so far.

Preview Of "Till The End: A Seven Day Devotional"

This is a sneak peak for the book "Till The End: A Seven Day Devotional" by Daniel Albertson, Jr. coming Late 2023.

Synopsis:

From the first sin in the garden to Jesus to The New Earth this devotional shows why we need Jesus and what is The New Earth that's coming.

Intro(For Till The End)

Dear, Reader

From the creation of the world to the fall in the garden to the sacrifices in the desert to the final sacrifice on the hill to The New Earth, this devotional will cover it all divided into seven days.

Summery:

Day 1: This is about how God created the world and who God is.

Day 2: This part is about the fall of humanity in the garden and the events that follow.

Day 3: This part is about the sacrifices that were supposed to clean God's people of sin but they continue to sin.

Day 4: This part is about Jesus and the blood he shed to we could all be healed.

Day 5: This part is about how to get to Jesus and further explains Day 4.

Day 6: This part is about the continued sinning of humanity and the signs of the coming new world.

Day 7: This last part is about the new world that is coming, how to get there, and the vents that lead up to it.

I hope you enjoy it and may The LORD speak to you.

με εκτιμιση,*

Daniel Albertson Jr.

*Greek For: "Me Ektimisi" Or Sincerely

About the Author

Zeno Aurilius is a political writer and journalist who writes about different current topics and problems.

About the Novel

This novel covers many topics from science to politics to culture and art. It is written in short essays that are either argumentative or discursive.

Notes: